Grand Solar Minimum

D.L.Bjorngjeld

Published by D. L. Bjorngjeld
Forest Lake, Minnesota

ISBN #978-0-9981110-5-6

Layout & Design by Todd Anderson

Printed in United States of America

What is a Solar Cycle?

The sun is a huge ball of hot, fluid, electrically charged plasma. It is the most important source of energy for all life on Earth. It is continually moving, turning, and generating a powerful magnetic field. This magnetic field goes through cycles, called solar cycles, every 11 years.

During the 11 year cycle, the sun's magnetic field flips and makes its North Pole and South Pole switch. These cycles affect activity on the surface of the sun, such as sunspots, which are caused by the sun's magnetic field.

Solar cycles can be tracked as the number of sunspots increase. At the beginning of a solar cycle, the sun has the least activity and is called the solar minimum.

Over the next few years, solar activity and sunspots increase. The peak at the middle of the solar cycle is called the solar maximum, the time period when the sun has the most activity and sunspots.

As the cycle comes to an end, it fades back to the solar minimum again, then a new cycle will begin. We are currently coming to the end of Solar Cycle 24 (SC 24), which has been a cycle of very low sunspot activity. SC 25 is predicted to have even less solar activity and will be the 4th consecutive cycle of diminishing activity.

The possibility of a Grand Solar Minimum is gaining evermore support. Even Nassau appears to be on board with their recent SC 25 headline prediction:

June 18, 2019 NASA

NASA Predicts Next Solar Cycle, SC 25, Will Be Lowest in 200 Years (Dalton Minimum Levels Could Be Reached).

What is a Grand Solar Minimum?

The next three or four solar cycles are projected to be lower solar activity than the current one that's ending. This indicates SC 25, 26 and 27 will be cycles with very little or no sunspot activity. Extended periods of inactivity like this, is the definition of a Grand Solar Minimum. This minimum will likely cause Global Cooling and, perhaps, a mini ice age.

Russian Scientists originally thought the minimum would happen slowly, but according to researchers at the Lebedev Physics Institute in Moscow, it's already happening. Sunspots and associated solar storms are already diminishing.

"Based on the picture that we are seeing currently, the Sun is moving quickly toward low activity cycles that will likely be reached by the winter of 2019-2020," the Solar Scientists said in a recent statement.

According to researchers in Moscow, the regions of high activity will disappear, solar radiation will drop to near zero, and sunspots will nearly disappear in the very near future.

"Finally, at the minimum point, the solar magnetic energy almost completely vanishes. The sun will continue on in this condition for several months, and up to a year. Then new fluxes of the magnetic field begin to float from the depth of the Sun to the surface, the first spots appear, and the solar cycle starts a new 11 year curve."

The Little Ice Age (LIA) was dominated by four consecutive solar minimums: The Wolf, Sporer, Maunder, and Dalton solar minima. All of which lengthened the cooling of the LIA to 450 years.

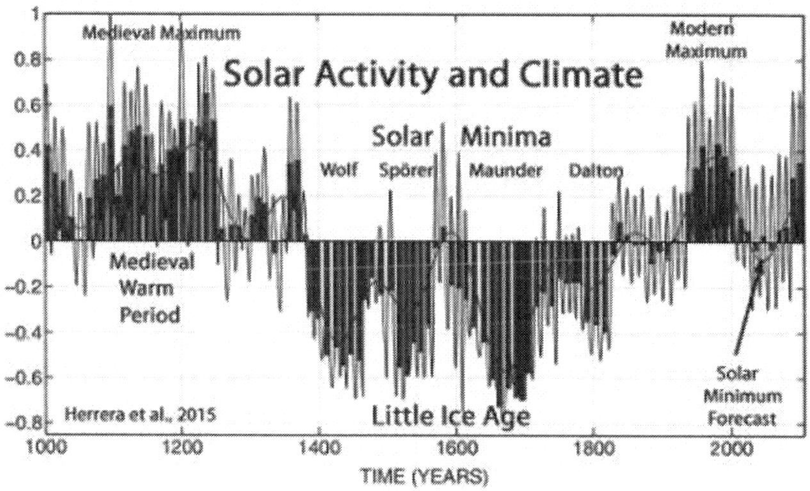

The **Wolf Minimum** was a period of low sunspot numbers in the 1300s.

The **Spörer Minimum** was a 90-year span of low solar activity, from about 1460 until 1550. It was named for the German astronomer Gustav Spörer.

The **Maunder Minimum** was from 1645 to 1715, during which sunspots became exceedingly rare. It was named for Astronomer E.W. Maunder.

The **Dalton Minimum** was a period of low solar activity, named after the English meteorologist John Dalton and lasted from about 1790 to 1830.

The Little Ice Age brought colder winters to Europe and North America. In the 17th-18th centuries, farms and villages in the Swiss Alps and in Norway were destroyed by encroaching glaciers. Canals and rivers in Great Britain and the Netherlands were froze, and many early settlers in North America reported exceptionally severe winters.

An Ice Age is triggered when summer temperatures in the northern hemisphere continually fail to rise above freezing for years. This means that winter snowfall doesn't melt, but instead builds up, compresses and, over time, starts to compact or glaciate into ice sheets.

Sunspots are somewhat cooler than the rest of the Sun, but when there are lots of sunspots, the Sun releases much more radiation. This greatly adds warming to Earth's climate. With fewer sunspots, the Earth's climate grows cooler over time.

High sunspot activity will make it warmer here on Earth. Low or no sunspot activity will cause cooling.

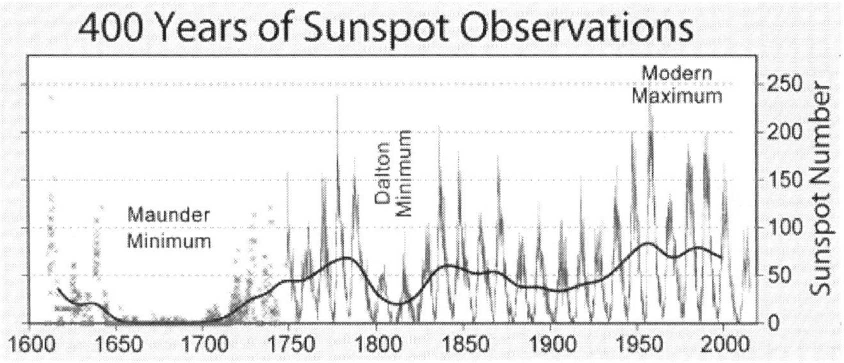

NASA is forecasting a possible return to the Dalton Minimum (1790-1830), *but gives no mention of the brutal cold, crop loss, famine, and death associated with it.*

Like the deeper Maunder and Spörer Minimums preceding it, the Dalton brought on a period of lower-than-average global temperatures. The Oberlach Weather Station in Germany, for example, experienced a 2°C decline in just 20 years, which devastated the country's food production. World-wide the climate became colder.

The "***Year Without a Summer***" also occurred during the Dalton Minimum in 1816. It was caused by a combination of already low temperatures plus the after-effects of the second largest volcanic eruption in 2,000 years. We are still currently in one of Earth's coldest periods and could face devastating cold if we enter a Grand Solar Minimum.

A Grand Solar Minimum is likely coming, and with it comes Global Cooling and maybe another mini Ice Age.

"Climate change continues (as it always has) however, there is verified scientific evidence that is pointing to a totally different result than the IPCC and Believers have been proclaiming."

Earth's Climate is a huge, immensely complex system. So far, however, there are strong indications *that we've ignored* a far greater factor in our climate than previously acknowledge. The factor that's been ignored is the influence that sunspots and solar activities have on our climate. We might well be ignoring this to our peril.

The frequency and intensity of sunspot activity has been shown to have a profound influence on Earth's weather systems, atmospheric temperature, ocean temperature, Gulf Stream flows and more. *It is also notable that the IPCC simply dismisses such influences as insignificant... A huge mistake.*

While governments and the United Nations have funneled Billions of dollars to the computer *modelers* to input various levels of CO2 in their models, *far too little attention has been given to the effect of the sun on earth's climate. At a minimum, in the interest of science if not our very survival, we need to remedy this.*

According to Prof. Masayuki Hyodo of the Research Centre for Inland Seas, Kobe University: *The United Nations' IPCC has discussed the impact that cloud cover and sunspots have on climate during their evaluations, but these phenomena have never been considered in their climate modeling or their predictions... All due to their insufficient understanding and inability to model them with accuracy.*

A Grand Solar Minimum and the Likely Effects on Climate.

If the predictions of the Russian Solar Physicists are accurate (and we will know in the next year or two, as opposed to the more than three decades of waiting for anything catastrophic to happen with the Warming Alarmism) we are in for a colder climate and unpleasant changes globally over the coming decades – perhaps until the 2050s – due to the onset of a new *Grand Solar Minimum.*

It is the conjunction of an 11 year cycle of declining sunspot numbers, with longer wave 100 year solar cycles. If this is so, the extreme weather conditions, severe cold waves, and more earthquakes, all could be predictors of what we'll be living with for the next 30–40 years, or longer. With that will likely come shortened growing seasons around the world, harvest failures, and famine in poor areas of the world.

A highly respected Solar Physicist studying solar cycles is Professor Valentina Zharkova, who teaches mathematics at Northumbria University in England. At a conference of the Global Warming Policy Foundation in October 2018, Zharkova presented findings that suggest, "***In 2020 a Grand Solar Minimum period could begin.***" (<u>In the past, predictions from her models have demonstrated **95%** accuracy.</u>)

Evidence, from independent scientists, suggest that we need to develop a better understanding of the sun and its effects on climate, if we are going to prepare for major climate catastrophes in coming years. ***(But the IPCC and Believers continue to assure us, "It's the CO2, Stupid.")***

Grand Solar Minimum: What If?

Climate events have been overly-dramatized by advocacy groups (particularly the political left and the United Nation's IPCC) contending that man-made global warming by the emission of CO2 is the cause. We've been inundated with proposals for <u>new taxes in the Hundreds of Billions of dollars</u>, taxes that we're told are needed to solve this problem." ***What if we are focused on the wrong problem?***

Recent research suggests that we have been too limited in our outlook and *are virtually ignoring some much greater influences in global weather than any man-made emissions.*

What is relevant to this discussion is the fact that *no climate model* used by the IPCC, or any of the hundreds of climate think tanks around the world, *are able to 'accurately model'* what is ***by far the greatest factor affecting our weather... The Sun.***

Currently, since 2018, we are experiencing a period of significant decline in solar activity. There is convincing evidence that this will bring us to what is called a Grand Solar Minimum, far more devastating than anything in our recent history.

According to Dr. Roy Spencer, "Clouds are the Earth's sun-shade, and if cloud cover changes for any reason, you will have global warming or global cooling." A Grand Solar Minimum will allow more Cosmic Rays to enter our atmosphere, having a direct effect on cloud cover.

The US Government's National Oceanic and Atmospheric Administration (NOAA) now tell us, "All weather on Earth, from the surface of the planet out into space, begins with the Sun." *(This is not what we've heard for the past three decades, when they were marching in stride with the "CO2 emissions are what is driving global warming" proclaimed by the IPCC and Believers.)*

What are the effects of less solar activity?

We will experience less solar energy or radiance reaching Earth. In addition to affecting cloud coverage globally, vital jet streams will be weakened and volcanic activity will be increased. We'll likely experience many more earthquakes and erratic unpredictable weather. The Jet Stream will shift south as it did in January, 2019 in North America, causing the record cold and snows across the region.

The Arctic blast that descended on a swath of the country stretching from the Rocky Mountains to New England continues, bringing record-breaking low temperatures, heavy snowfall in some Northeast areas, and school closings in the Mid-South. December, 2019.

The current solar cycle, designated by NASA as number SC 24 peaked in early 2014 before starting its measurable decline in annual sunspot activity. A Grand Minimum is predicted to take place in 2020 and could last for a few decades. Some predict it will cause a new "mini Ice Age."

The subject of sunspots is complex and vastly under-researched, as we have focused instead on man-made changes to our climate with simplistic computer models. *If the coming winter in the Northern Hemisphere is anything like the past one, it should prompt us to take this solar component of our climatic influences seriously.*

By refusing to promote vigorous new research, we run a real risk, in coming years, of being unprepared for dramatic cold resulting in harvest failures, and at a time when most governments have decided to eliminate emergency public grain reserves, and our food supply operates on a "just-in-time" system.

Science is about discovering the *truth*, wherever it leads and however controversial.

ANOTHER CLIMATE SCIENTIST WITH IMPECCABLE CREDENTIALS BREAKS RANKS WITH THE IPCC, SAYING: "OUR MODELS ARE SIMPLY MOCKERIES OF THE REAL WORLD"

Dr. Mototaka Nakamura received his Doctorate of Science from the Massachusetts Institute of Technology (MIT), and for nearly 25 years specialized in abnormal weather and climate change at prestigious institutions that included MIT, Georgia Institute of Technology, NASA's Jet Propulsion Laboratory, California Institute of Technology, and Duke University.

In his book titled, "The Global Warming Hypothesis is an Unproven Hypothesis," Dr. Nakamura explains why the data underpinning the Global Warmist science is doctored, untrustworthy, and cannot be verified or relied on.

From 1990 to 2014, Nakamura worked on cloud dynamics and forces mixing atmospheric and ocean flows on planetary scales. His home-bases were MIT (for a Doctor of Science in Meteorology), Georgia Institute of Technology, Goddard Space Flight Center, and the Japan Agency for Marine-Earth Science and Technology.

"Today's 'global warming science' is akin to an upside down pyramid which is built on the work of a few climate modelers. These 'few' claim to have demonstrated that human-derived CO2 emissions are the cause of recently rising temperatures and have then simply projected that forward. Every climate researcher thereafter has taken the results of these original models as a given. ***Merely testing their validity is regarded as heresy.***

With Nakamura, we have a highly qualified and experienced climate modeler with impeccable credentials ***rejecting the unscientific basis of the climate crisis claims***. But he's up against a big army. The activists seem to be winning at the moment, and they're now being fronted by scared, crying children; an unstoppable combination. One that's tricky to discredit without looking like a heartless bastard.

NAKAMURA ON DATA FALSIFICATION

When arguing against global warming, the hardest thing is *convincing people of data falsification, namely temperature doctoring*. If you don't pick your words carefully, or get your tone right, then it's very easy to sound like a conspiracy crank.

Now we have Nakamura, one of their own. He has accused the "climate scientists" of data falsification in the form adjusting historical temperature to *inflate today's subtle warming trend*.

Additionally, the climate models grossly simplify the way climate works. As well as ignoring the sun's effect, they drastically simplify ***large and small-scale*** **ocean dynamics**, ignore **changes that generate** ***clouds*** (cloud cover is one of the key factors determining whether we have global warming or cooling), and ***water vapor***.

NAKAMURA ON CO2

"Earth's climate system is far more complex than an absurdly simple system simulated by the 'models' that have been used for climate projections.

"The dynamics of the atmosphere and oceans are absolutely critical facets of the climate system if one hopes to ever make any meaningful prediction of climate variations.

"Solar input is modeled as a 'never changing quantity,' which is absurd.

NAKAMURA ON SOLAR FORCING (SOLAR INFLUENCE ON EARTH'S CLIMATE)

Solar output is not constant, which is what the IPCC insists. During solar minima, like the one we're entering now, the sun's magnetic field weakens and the outward pressure of the solar wind decreases.

This allows more Cosmic Rays from deep space to penetrate our planet's atmosphere. These CRs have been found to affect clouds, and clouds are a crucial player earth's climate. As Roy Spencer, PhD, writes: *"Clouds serve as Earth's sunshade. If cloud cover changes for any reason you will have global warming or global cooling."*

Cold times appear to be returning, as a result of ***historically low solar activity.*** NASA has warned that this next solar cycle (Cycle 25) will be "the weakest of the past 200 years":

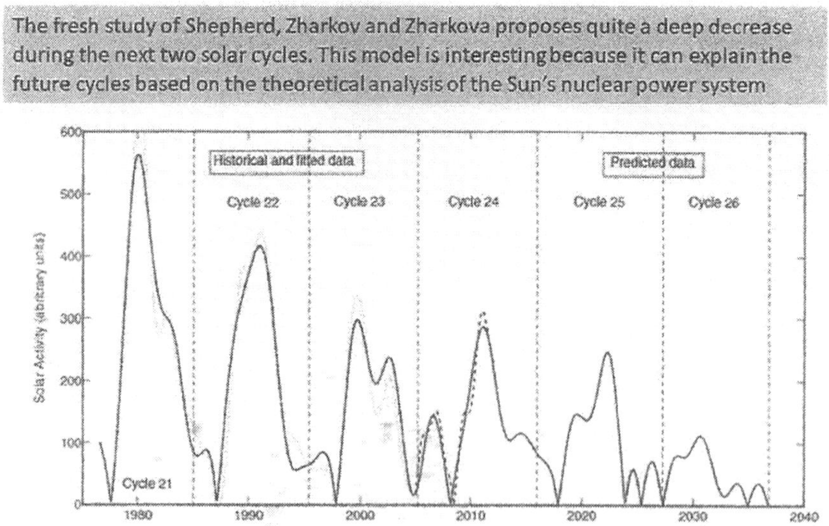

The Maunder minimum, AD 1645–1715, is the most recent Grand Solar Minimum, and the only one covered by telescopic sunspot observations.

The current cycle, SC 24, is the lowest sunspot cycle in 100 years and the third in a trend of diminishing sunspot cycles. Solar physicists expect cycle 25 to be even less active and expect the sun to move into a new solar minimum, one comparable with the Dalton Minimum. Studying such a minimum with available modern instruments could potentially answer a lot of the serious questions surrounding the influence of the sun on our climate.

Previously, researchers used the number of sunspots to represent the activity of the solar magnetic field. New technology takes advantage of direct observations of magnetic fields emerging on the surface of the Sun – data which has only existed for the last four solar cycles.

SC 25 may likely be a mere stop-off on our descent into the next Grand Solar Minimum. The excellent observation and analysis of Dr. Valentia Zharkova suggests that all four of the Sun's magnetic fields will go out of phase in 2020, which strongly suggests that we're headed into a very cold climate.

Professor Zharkova gave a presentation of her Climate and the Solar Magnetic Field Hypothesis at the Global Warming Policy Foundation in October, 2018. The information she unveiled should wake up the Warmists.

Zharkova was one of two scientists to correctly predict solar cycle 24 would be weaker than cycle 23. In fact, only 2 out of 120+ climate models predicted this correctly.

The last time we had a GSM (the Maunder Minimum) only two magnetic fields of the sun went out of phase. This time, all four magnetic fields are going out of phase.

Professor Zharkova's model has shown 95% accuracy and now is suggesting that a Grand Solar Minimum and global cooling will be headed our way beginning in 2020.

How cold might it get? Based on the study of previous sunspot reductions proceeding a grand minimum period, we will see a 7% reduction in the Sun's heat and light. This will likely result in bitterly cold winters and shorter summers in North America and parts of Europe.

Meanwhile, the United Nation's International Panel on Climate Change (IPCC) has countered all of this new information from the last couple of years, which shows a strong potential of Climate Cooling; AND the IPCC's poor reputation in their ability to model Earth's climate; AND the evidence of doctoring historical records to prop up their claims; with them producing and ***proclaiming an ever more frantic barrage of new, more extreme, and more fraudulent alarmism.*** The following is a good example of their frenzied alarmism and, many times, outright lying:

"We could be losing the race against climate change, new UN report says."

Sea walls all across the globe are being rendered useless by increasingly volatile seas all due to sea level rises.

Scientists behind a study of the links between oceans, glaciers, ice caps and the climate delivered a stark warning to the world: ***slash emissions or watch cities vanish under rising seas.***

The U.N. backed panel of experts has found that radical action may yet avert outcomes of global warming. The study was clear that allowing CO2 emissions to continue on their upward path would upset the balance of the great geophysical systems governing oceans and the ice regions of the earth so profoundly that no one would be untouched.

Are the sea levels continuing to rise? Yes they are... At an average rate of 2 mm/year, as they have since we've been keeping records. (A rise of 2 mm/year <u>over ***12 years*** equals one inch of sea level rise.)</u>

Sea levels are not, "rising at 10 times the rate seen in the 20th century."

We are not, "releasing enormous quantities of CO2 which will *unleash feedback loops driving ever faster warming."*

We are not, "causing oceans to get hotter, bringing on what are known as Marine Heat Waves that will damage ecosystems."

We are not, "heating the oceans and starving the upper layers of oxygen, suffocating marine life, creating growing dead zones, and disrupting the circulation of ocean currents."

If emissions do continue rising – as they have for the past 70 years – ***it will not*** "accelerate impacts so rapidly that they will overwhelm society's ability to cope."

Choose a descriptive word for the proclamations of the IPCC, their Believers and fellow alarmists:

Absurdity	Fabricated
Poppycock	Distorted
Horse-Feathers	Preposterous
Balderdash	Ridiculous
Baloney	Hogwash
Twaddle	Prattle

I'm afraid that we might be wishing for a bit of warming, if Dr. Zharkova is correct in her prediction of a coming Grand Solar Minimum.

My thoughts...

I'll take a little warming, like that of +0.8°C during the 20th century.

If a 'Grand Solar Minimum' happens and influences climate as it did in the 15th–19th centuries, we'll see lots of this. AND, what is the IPCC going to do about stopping it?

Made in the USA
Monee, IL
13 September 2026

62979534R10015